GERMS AND DISEASE

WHAT ARE FUNGI AND MOLDS?

By Kathryn Kroe

Cavendish Square
New York

Published in 2023 by Cavendish Square Publishing, LLC
29 East 21st Street, New York, NY 10010

Website: cavendishsq.com

Library of Congress Cataloging-in-Publication Data

Names: Kroe, Kathryn, author.
Title: What are fungi and molds? / Kathryn Kroe.
Description: First. | New York : Cavendish Square Publishing, [2023] |
Series: Germs and disease | Includes index.
Identifiers: LCCN 2021059674 | ISBN 9781502665430 (library binding) | ISBN 9781502665423 (paperback) | ISBN 9781502665447 (ebook)
Subjects: LCSH: Fungi—Juvenile literature. | Molds (Fungi)—Juvenile literature.
Classification: LCC QK603.5 .K76 2023 | DDC 579.5—dc23/eng/20211213
LC record available at https://lccn.loc.gov/2021059674

Editor: Kate Kroe
Copyeditor: Jane Katirgis
Designer: Andrea Davison-Bartolotta

The photographs in this book are used by permission and through the courtesy of: Cover Graeme Pearce/Shutterstock.com; series art (backgrounds) CHIARI VFX/Shutterstock.com; p. 5 Ksu Shachmeister/Shutterstock.com; p. 6 Vladimir Mulder/Shutterstock.com; p. 8 Moving Moment/Shutterstock.com; p. 9 Designua/Shutterstock.com; p. 10 Jirawan muangnak/Shutterstock.com; p. 12 Leonardo Mercon/Shutterstock.com; p. 13 Kcuxen/Shutterstock.com; p. 16 Konoplytska/Shutterstock.com; p. 19 AlexeiLogvinovich/Shutterstock.com; p. 20 Andrii Repetii/Shutterstock.com; p. 22 Christian Delbert/Shutterstock.com; p. 24 Dan Gabriel Atanasie/Shutterstock.com; p. 26 WAYHOME studio/Shutterstock.com; p. 28 JACGS/Shutterstock.com; p. 30 Dejan Stanic Micko/Shutterstock.com; p. 31 Victoria 1/Shutterstock.com; p. 34 BAZA Production/Shutterstock.com; p. 36 Syda Productions/Shutterstock.com; p. 37 victoras/Shutterstock.com; p. 39 Simone Hogan/Shutterstock.com; p. 40 M-Production/Shutterstock.com.

CPSIA compliance information: Batch #CSCSQ23: For further information contact Cavendish Square Publishing LLC, New York, New York, at 1-877-980-4450.

Printed in the United States of America

CONTENTS

INTRODUCTION

Have you ever opened a container of berries to find several of them covered in fuzz? Or maybe, on a hot summer day, you opened a bag of bread to find green spots all over the slices. In both cases, the food is ruined and unsafe to eat. The culprit of your ruined snack is fungus.

Still, in the same kitchen where you might find fungi ruining your food, they are also essential to so many other items. Maybe you have mushrooms or yeast, both of which are members of the same kingdom as the fuzzy, green stuff you avoid.

As with the organisms in most kingdoms, the diversity among those in the kingdom Fungi is great. Think about the kingdom Animalia. Both humans and turtles belong to it, and they are about as different as can be. Certain characteristics are required to be classified into a kingdom, but beyond those, the organisms can be unique. Mold, for example, is a specific type of fungus that grows in multicellular structures called hyphae.

It is easy to be grossed out by fungi or worry about their impact on our health. Fungal infections can affect us both internally and externally, especially in people with weakened **immune systems**. While these concerns are valid, fungi are still important to our everyday lives. Like bacteria, fungi exist almost everywhere on the planet. They are hardy organisms that thrive in places you might not expect. Also like bacteria, they are essential in the breakdown of dead and decaying matter. One important group of fungi that does this is known as saprophytes. These fungi use decaying matter for nourishment, and without them breaking down organic matter throughout the environment, life on Earth would be quite different. Without fungi, the world we know could not exist.

These big mushrooms are part of a fungus and one of the varieties of mushrooms that is safe to eat.

Fungi don't need light to create food. They feed off outside sources, so they can thrive in a place such as this cave.

CHAPTER 1

HOW DO WE CLASSIFY FUNGI AND MOLDS?

While a fungus might remind us of a plant, there are a few important differences. Fungi are nonflowering organisms that do not make their own food. Instead, they absorb food from outside sources. Another important distinction is their cellular makeup. While both fungi and plant cells have a cell wall, the wall on a fungus cell contains chitin rather than cellulose. Chitin is the same material that makes up an insect's exoskeleton.

FUNGI VS. PLANTS

Fungi and plants have several things in common, but a few important characteristics set the kingdoms apart.

Because fungi are not plants, they don't require light to thrive. In fact, they often live in dark places such as under piles of wood, in basements, or in caves. This can make them hard to notice until they have greatly spread.

Mold is a type of fungus. One example of mold is the fungus with a fuzzy texture that we often see develop on our food products. Mold prefers warm, moist places to thrive. For this reason, you may notice mold in a place such as your shower. While we usually picture mold as undesirable, it is also used purposely in some food, including many expensive cheeses.

The white outer shell on this Brie cheese is actually mold!

Fungi Anatomy

The body of a fungus is not like the body of a human or an animal, or even like that of a plant. As humans, we are fairly compact creatures that take up a fixed amount of space. Some fungi can spread for enormous distances. A single fungus discovered in a Seattle, Washington, forest in 1992 had a body that spread for 2.5 square miles (6.5 square kilometers). Overall, it weighed nearly 1,000 tons (907 metric tons). Some scientists believe it is the largest creature ever to have lived on this planet.

Although many fungi and molds are huge, they have very simple structures, especially when compared to other living creatures, such as humans. Our bodies are made up of many different parts and organs. We have hands, feet, legs, arms, a heart, a brain, and more. Fungi possess only four main parts: hyphae, the mycelium, a fruiting body, and spores.

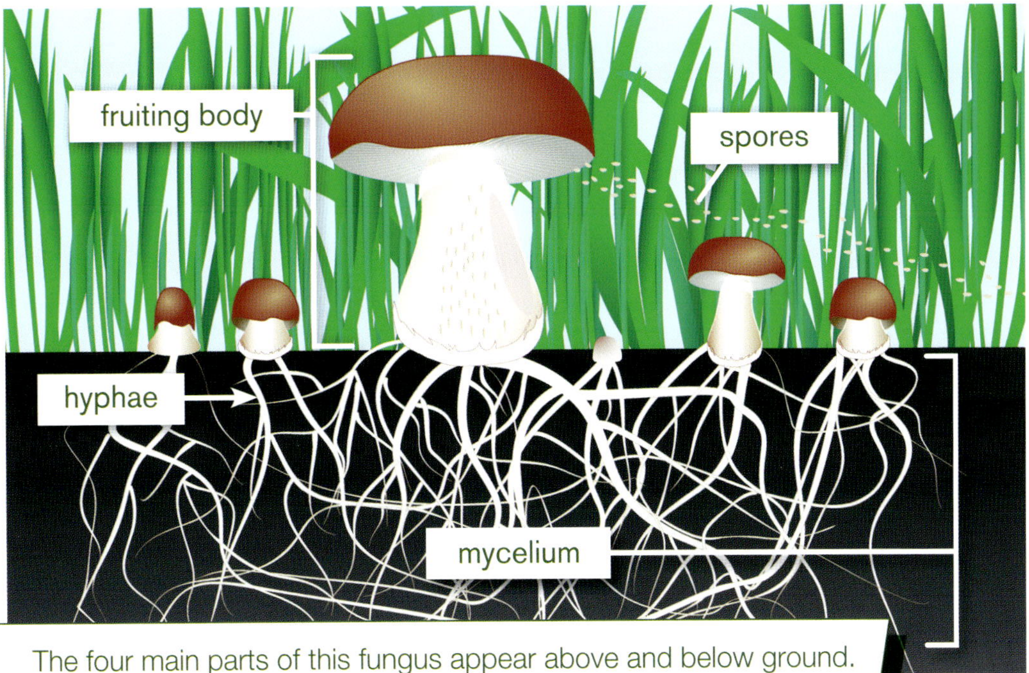

The four main parts of this fungus appear above and below ground.

HYPHAE

The body of a fungus is made up of tiny, hairlike strands called hyphae. These strands can be very thin and difficult to see without using a microscope. Some species of fungus have hyphae so tiny that it would take 50,000 of them laid together to make one inch (2.5 centimeters). The huge Seattle fungus was not one huge, visible mass lying on the ground. Most of it would have been invisible to passersby. That is because most of the hyphae were underground.

Whether aboveground or underground, hyphae spread out in a network of twisted strands, the way roots grow. Roots, however, tend to grow in three dimensions—up, down, and sideways—while the hyphae of a fungus tends to spread out more in the shape of a flat mat.

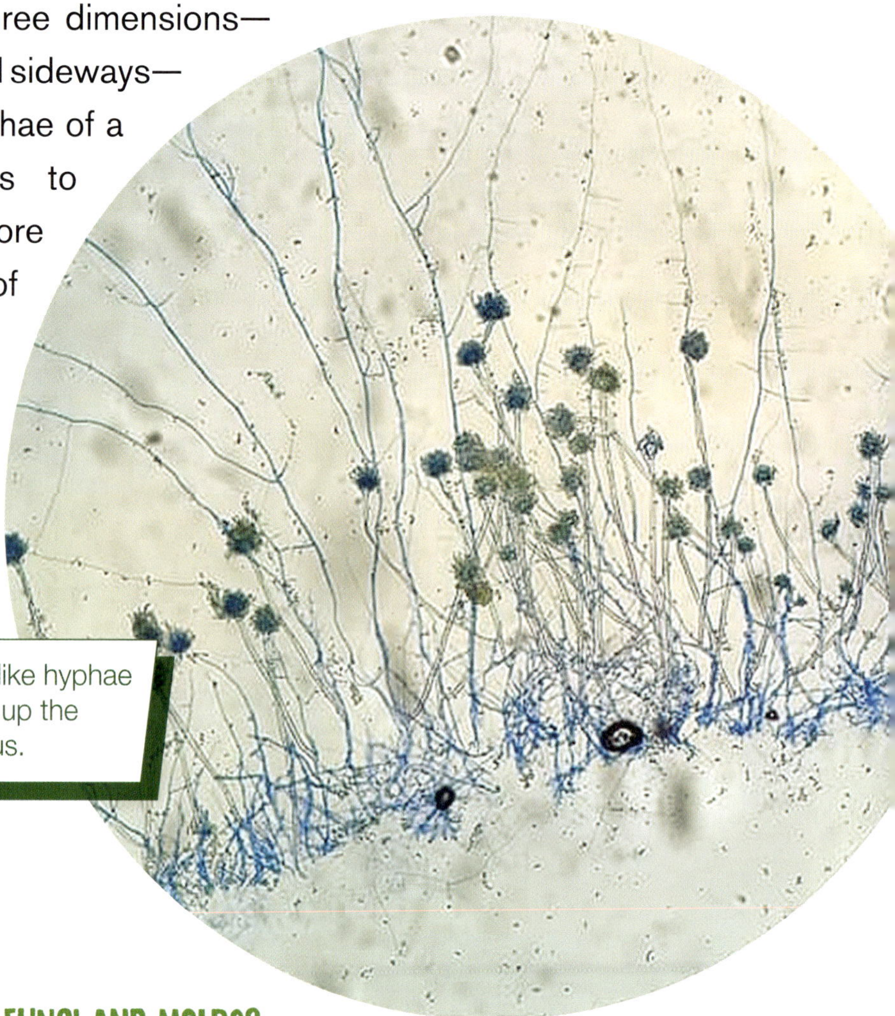

These tiny, hairlike hyphae are what make up the body of a fungus.

Think of the hyphae as being like fibers in a square rug. The hyphae that lie in the inner section are responsible for transporting food and water to the fungus. Because they must absorb water and nutrients, these hyphae have many small openings. The hyphae that lie on the outer edges instead are covered with protective walls. Again think of a rug, where the edge is made up of sturdy, tough fibers, while the inside fibers have holes or gaps.

Mycelium

Just as individual fibers put together make up a rug, each hypha (singular for hyphae) helps make up a mycelium. Mycelium is a term used to describe the entire mat of tangled hyphae fibers. Because the hyphae carry food and water to the fungus and a fungus does not need light, the mycelium functions like the roots, stems, and leaves of a plant.

The mycelium enables a fungus to spread to other areas. If you have ever worked in the garden, you might have helped someone separate plants that grow together. For many species, the plant can be broken off, with roots attached, and planted in another spot where it will grow into a whole new plant. The same is true for many fungi. For these species, a patch of the mycelium can be broken off and placed in a new area, where it will grow into another fungus. Often, commercial growers of mushrooms use this technique to increase their yield.

Fruiting Bodies

Mushrooms are one example of a fruiting body. You might think that a mushroom is an entire fungus, but it is only one small part of the entire growth. Underneath mushrooms are the hyphae that make up the mycelium, which takes up much more space. The mushroom serves as a fruit of the fungus. Just as a single apple is only one small part of a big tree, a single mushroom is only one fruiting body of a larger fungus.

It is incredible to think about how big an average garden fungus grows. When outside, look for mushrooms. (Do not pick or eat any, though! A number of mushrooms growing in the wild are very poisonous.) Observe how they grow. Sometimes they can be found growing in a circle or in clusters. Now, visualize that these fruiting bodies are attached to hyphae that could radiate for miles around where the mushroom was found.

Just as fruit often contains seeds, fruiting bodies contain spores that are like the seeds of the fungus. When dropped in an appropriate environment, the spores can develop into new fungi. Fruiting bodies may produce tens of thousands of spores. Some fungi are able to shoot them out like a water pistol.

In a mushroom, spores are produced by the gills. The cap is the part of the mushroom that looks like the top of an umbrella. A stalk holds up the cap and the gills. It is not easy to see the spores contained within the gills, but try placing a cap on a white paper towel under a cup. Leave it there overnight. In the morning

The spores, which are like the seeds of a mushroom, are contained between these gills under the cap.

when you pick up the cap, you should see tiny specks on the paper towel. These are the spores.

YEASTS AND SLIME MOLDS

There are many different types of fungi. Just as mold is a type of fungus that includes numerous species, yeasts and slime molds are two other types of fungi. Yeast is a single-celled, microscopic fungus. It exists in the wild, occurring on plants and in soil. Certain kinds are also **cultivated** in a factory setting for use in making bread, wine, beer, and other foods and drinks. Yeast feeds on sugar, nitrogen, minerals, and water. When used in making bread, yeast is fed a small amount of one or more of these things in a process called proofing. As the yeast feeds, it gives off carbon dioxide gas. When bread is baked, the gas is trapped in little holes in the dough, which makes the bread rise into a fluffy loaf.

Have you ever made your own bread? When it gets big and fluffy like this, it's actually the fungus yeast doing its job!

GROW YOUR OWN MOLD!

While we know mold can grow and thrive in many conditions, it commonly thrives on food products in our homes. With the help of an adult, you can create your own science experiment to answer the question, "Which food will grow the most mold?" Follow the steps to complete the experiment.

1. Select some different foods to test. Things such as bread or fruit are great options, but try whatever you'd like!
2. Cut each food into six samples, and place them in individual plastic bags. You will have six bags for each different food type you are testing. Label the bags with the type of food.
3. Leave three bags of each food dry. Add the label DRY to those bags.
4. Wet the food samples in the other three bags of each food. Add the label WET to those bags.
5. Place a dry and wet sample of each food into the refrigerator.
6. Place a dry and wet sample of each food into a sunny, warm location.
7. Place a dry and wet sample of each food into a dark location, such as a closet.
8. Make predictions, or hypothesize, about which foods will grow the most mold.
9. Monitor your samples over time and note the changes.
10. Ask yourself if your hypothesis was correct.

Slime molds, as their name implies, have a slimy texture and are much less appetizing than the fungi used in preparing foods. Scientists usually put these in a separate category from other molds because of their somewhat different appearance and behavior. These molds move along a surface by changing their shape, absorbing nutrients and water using special cells. Once the surface is dry, the slime actually transforms into a collection of fruiting bodies that then create and disperse spores. These travel through the air. Wherever they come upon moisture, new slime molds develop.

It is easy to see many fungi as a nuisance or as potential poisons, as many types of them indeed are. There are many cases in which human interaction with fungi can cause ill health effects and even death. However, the latest scientific estimates are that there are as many as 5 million different species of fungi on Earth, and only a few hundred of these are known to sicken human beings. The remaining species live with us in harmony. It is important to understand when fungi are harmful but not fear the organisms as a whole.

WHAT DO YOU THINK?

1. Why do some people confuse fungi with plants?
2. Why is it important not to eat mushrooms you see in the wild?
3. Have you ever eaten any of the examples of edible fungi or mold mentioned in this chapter? Did you know what it was when you ate it?

The white you see on the fruit is the hyphae of the mildew fungus. It feeds off the nutrients in the fruit, killing that section in the process.

CHAPTER 2

WHEN ARE FUNGI HARMFUL?

Fungi that live off plants or animals are called parasites. Not all fungi are parasites, but many of those that are can be harmful. They nourish themselves from the host plant or animal and do not provide anything in return. One such parasite is called mildew. Other fungi leave behind poisons they produce when they feed. If humans **ingest** these poisons, they could become ill.

MALICIOUS MILDEW

Mildew refers to several different parasitic fungi that attack plants and things made from animal and plant products, such as books. Books are made from paper, which in turn is made from tree pulp. Mildew feeds off the pulp.

While the words "mildew" and "mold" sometimes are used interchangeably, mold usually has a color to it and can be fuzzy, while mildew generally is white to off-white and not fuzzy. Like mold, mildew grows in damp places. There are many kinds of mildew. Two major types are powdery mildew and downy mildew.

You may have seen powdery mildew on fruits or vegetables. The visible white blotches actually are the hyphae of the fungus. The hyphae branch out across the plant to absorb its nutrients. Often, in the process of feeding on the plant, the mildew will kill the sections that it has fed off. An apple covered with mildew, for example, could eventually rot even while still on the tree. Other plants commonly affected by powdery mildew include cherries, gooseberries, grapes, peaches, peas, and roses.

Downy mildew frequently is visible as yellow splotches on the surface of leaves or fruits. As the mildew feeds, it produces spore-like structures on the bottoms of the leaves, which in turn may produce even more spores. When these spores grow and develop, they send out hyphae that can extend into the interior portions of the plant. The plant becomes infected with the fungi and may die, unless the mildew is controlled.

Fungal Poisons

Some fungi produce poisonous chemicals as they feed. These chemicals are called **mycotoxins**. The term comes from the Greek word for "fungus," *myco*. As some fungi feed on animals, plants, or items made from animal or plant materials, they secrete substances that help the fungi to absorb nutrients. In humans, our digestive **enzymes** are located internally in our saliva, stomach, and intestinal tract. Fungi do not have stomachs so they must prepare their food for consumption outside of their bodies. The chemicals they release to do this can be poisonous to plants and humans.

Mycotoxins can form on moldy foods, especially peanuts, corn, rice, and wheat. People or animals that consume these affected foods can become sick. Sometimes livestock fed contaminated grains can then sicken humans who consume meat and dairy products made from the same livestock. **Aflatoxin** is the most common kind of mycotoxin. It is extremely dangerous. Studies show that it is the strongest cancer-causing agent to be identified to date. Two other mycotoxins are trichothecene and zearalenone. They can damage the intestines and other parts of the body when consumed.

Some studies now suggest that even inhaling mycotoxins can make a person ill. A black mold called *Stachybotrys chartarum*, commonly found on damp drywall, has been found to produce mycotoxins that appear to be harmful when inhaled.

Black mold is a dangerous mold that produces mycotoxins. If you see it in your home, it is important to get a professional to clear the mold in the area.

Without proper maintenance, air conditioning systems can be a breeding ground for mold. This can be especially dangerous because the system can spread the mold once it grows.

Many spaces we spend time in have the potential to be a breeding ground for mold and mildew. Our cars, our laundry rooms, and even our gym bags can provide hospitable environments for these fungi to grow if we are not careful. Besides looking and often smelling gross, molds can cause serious health issues. We need to be sure to keep things clean and dry whenever possible to prevent these germs from spreading.

WHAT DO YOU THINK?

1. Why are only some fungi considered parasitic?
2. If fungi leave mycotoxins on the source they are absorbing nutrients from, how can they make humans sick?
3. What are some other places mold and mildew might thrive?

THE LAUNDRY ROOM—CLEAN OR DIRTY?

The laundry room is where we take our clothes to clean them, so why is it at such high risk for being dirty? In this case, when we say dirty, we are talking about fungi. If we aren't careful, the laundry room can become like a five-star hotel for fungi.

Washers and dryers both involve moisture and heat, two factors that mold and mildew love. Think about your washer when you remove the clothes. If you don't allow the washer to dry out, that moisture sits in crevices like the seal and encourages mildew to grow. Therefore, it is always recommended to leave the washer door open after a load of laundry is complete. When a washing machine runs, water goes through hoses to travel in and out of the machine. If these hoses aren't sealed properly, especially if they are against a wall, the leaking water can encourage dangerous black mold to grow.

While most dryers don't have a water connection, the wet clothes and warm air create moisture. This moisture is released through a vent as the clothes dry. If this vent isn't properly installed, that moisture can stay in the dryer or laundry room and create mold and mildew too.

Your laundry room should be a clean place to clean your clothes. Being aware of where fungi can thrive can stop it from becoming a problem.

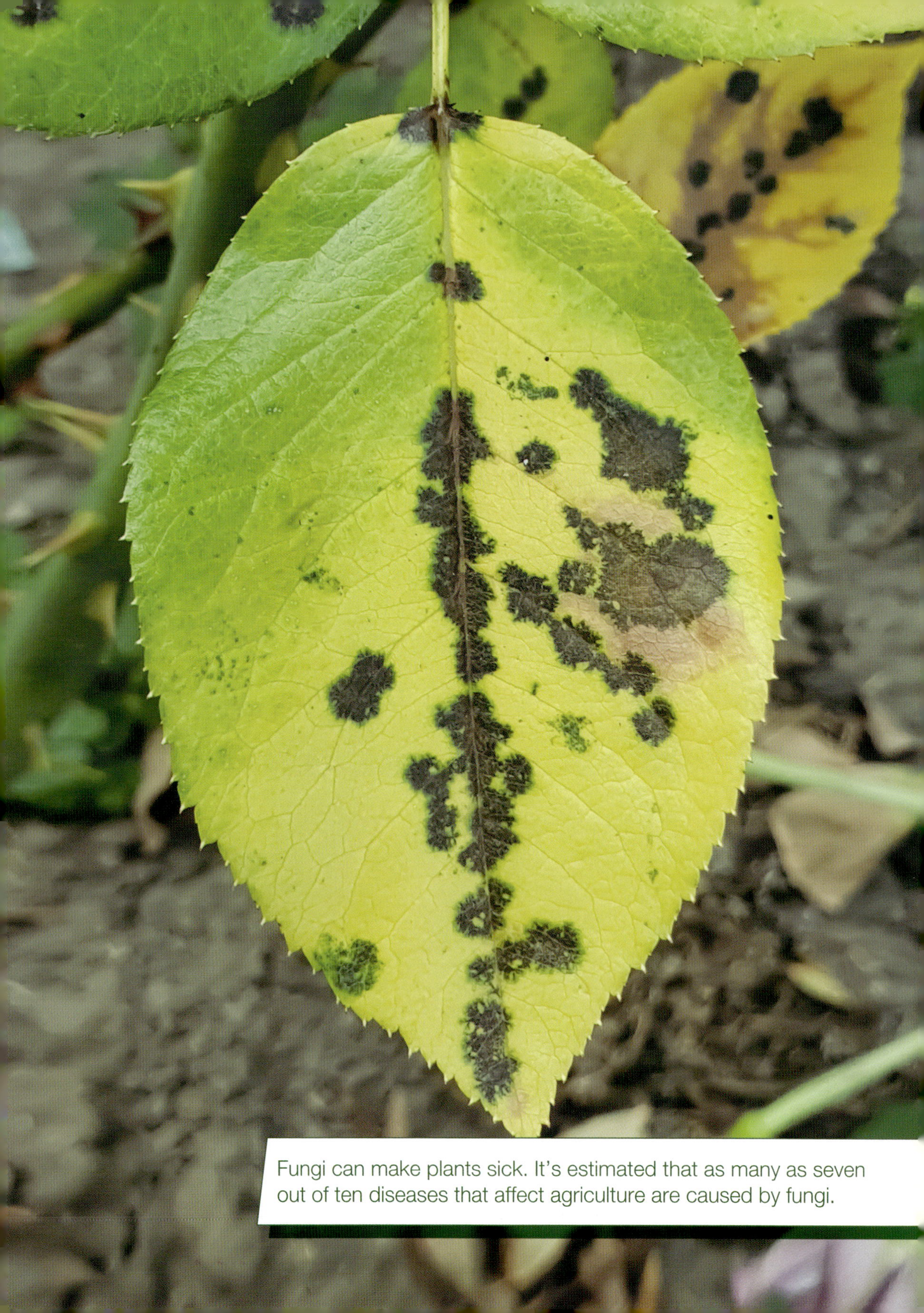

Fungi can make plants sick. It's estimated that as many as seven out of ten diseases that affect agriculture are caused by fungi.

CHAPTER 3

WHAT DISEASES CAN FUNGI CAUSE?

When we encounter harmful fungi, they can cause diseases, but human diseases are not the only concern when it comes to fungi. Like other germs, fungi can also affect plant life. Often, this effect is devastating to agriculture. It not only hurts the plants that die, but can also severely impact the livelihood and health of those who depend on those plants.

Besides the loss of plants we depend on, harmful fungi can affect us in several ways, ranging from allergies to serious **systemic diseases**. Some are only inconvenient or uncomfortable, while others can be life-threatening, especially for those who are **immunocompromised**. Human health problems caused by fungi generally are divided into three different categories: allergies, toxic reactions, and fungal infections.

Seasonal Allergies

Many people have experienced an allergic reaction at some point in their lives. One of the most common allergic reactions is kown as **hay fever**. It often happens during the spring or fall, the times when plants release a lot of pollen into the air. Like pollen, mold and mildew also are released into the air. On a sunny day, you may even be able to see what looks like dust flying around in the air inside your home. Some of those particles are probably mold spores.

Pollen is not the only thing that can cause our eyes to itch and noses to run. Allergies can also be brought on by spores in the air.

Not everyone is bothered by the presence of the spores. Many people, however, are sensitive to the proteins that make up the spores. When they breathe them in, their body's immune system reacts in way that is similar to a hay fever reaction, causing sneezing, runny noses, headaches, and other symptoms. That is why health organizations track not only the amount of pollen in the air at certain times of the year, but also the amount of mold spores.

TOXIC REACTIONS

People and animals can suffer, or even die, from eating fungal toxins in poisonous mushrooms or food that has been contaminated. Fortunately, most fungus-contaminated foods let us know that they are bad to eat because they smell or look awful. An orange covered with mold, for example, is not going to being very appetizing or healthy to eat. That is why it is important to shop carefully for fresh, uncontaminated foods. Avoid those that look spoiled or rotten. Spoiled foods often attract molds and mildews.

Store-bought mushrooms—the kind you find on pizza and in other foods—are safe to eat because they are carefully grown on farms that focus only on edible mushroom varieties. There are many mushrooms in the wild, however, that are extremely poisonous. Just one tiny bite from such a mushroom's cap can lead to severe illness and even death.

Many poisonous mushrooms come from a genus called *Amanita*. Many of these mushrooms can be found in parks and gardens, particularly after heavy rain. They can look like a common button mushroom, only with a narrower stalk and cap. Because poisonous mushrooms often look like edible fungi, it is very important to never pick a wild mushroom and eat it without the help of a trained mushroom expert.

While we may recognize this fungus as a mushroom, there are many kinds of mushrooms. It is important to never eat a mushroom growing in your yard or in the wild as it could be poisonous. Stick to the mushrooms you buy at the store!

FUNGAL INFECTIONS

Fungal infections, also called mycoses, are diseases caused by fungi. A fungus invades living parts of the body, which results in an infection. Only about 200 of the different species of fungi infect humans. Of these, only a few dozen cause most diseases.

In general, **pathogens** such as fungi are either considered primary pathogens or opportunistic pathogens. Primary pathogens can hurt anyone. Opportunistic fungi tend to affect people who have a weakened immune system, such as people already suffering from another disease, such as cancer or **AIDS**.

Fungal infections are grouped into four categories, depending on the extent of the disease and what part of the body they impact. The four groups are superficial mycoses, cutaneous mycoses, subcutaneous mycoses, and systemic mycoses.

Superficial Mycoses

Superficial mycoses are fungal infections of the hair and the outermost surface of the skin. While not life-threatening, they can cause discomfort. Most of these infections occur in tropical regions throughout the world where the weather frequently is warm and conditions are moist and humid.

One such infection, *Tinea versicolor*, is caused by uncontrolled growth of a fungus that is commonly present on human skin in small amounts. A little bit of the fungus causes no symptoms, but when it gets out of hand, an infected person may develop scaly patches on the skin that can develop into sores. Antifungal creams usually kill the excess fungi and get rid of the problem.

Tinea nigra is another superficial mycosis. It is caused by a fungus that usually infects the feet and the palms of the hands. These areas tend to release a lot of moisture. Think about when you have been active for a while, perhaps while exercising, hiking, or performing some other physical activity. When you remove your shoes, your feet may feel warm and moist. If you do not change

Sweaty feet are at higher risk for developing superficial or cutaneous mycoses. If your feet are warm and sweaty, it is important to remove shoes and socks so you can wash and dry your feet as soon as possible.

your socks and wash your feet, both the socks and your feet could become a perfect breeding ground for a fungus.

Cutaneous Mycoses

Fungal infections that fall under this heading affect more layers of the skin than a superficial infection. That means they go deeper than just the skin's outer surface. Cutaneous mycoses are not life-threatening but can worsen and become chronic, or more long term, if left untreated.

Athlete's foot is by far the most common cutaneous mycosis. You may even have had it yourself or know of someone who did. A fungus that attacks the feet causes it. Usually, athlete's foot spreads when a person steps on a surface that has been contaminated with the fungus, such as a shower floor.

Candidiasis is another common fungal infection of the skin. It is caused by an overgrowth of yeast, known as *Candida*, that everyone has in small amounts in the nose, mouth, and other areas. As the yeast feeds on the skin, it can damage it, causing lesions and pus-filled infections. Candidiasis may become quite serious if left untreated, as it can spread beyond the skin and into other parts of the body. For all fungal diseases, it is important to get early treatment before a minor problem develops into something worse.

Candidiasis is a common fungal infection that nursing babies can get in their mouths. The telltale symptom of a white tongue is a sign of an overgrowth of yeast in the condition commonly called thrush.

Subcutaneous Mycoses

"Subcutaneous" means "below the skin," so subcutaneous mycoses are fungal infections that can affect the skin, tissues below the skin, and bones. Usually, the fungi that cause these conditions live in soil and on decaying vegetation. Victims contract the fungus through an injury, such as walking barefoot on contaminated soil and cutting the bottom of a foot on something sharp, like the edge of a rock.

These kinds of infections are more common in tropical and subtropical areas, such as India, Africa, and Latin America. People who work outdoors, such as farmers, gardeners, and nursery workers, are at greatest risk. People often buy tea or coffee grown in one of these tropical or subtropical regions. Some buyers of these products in the United States and in other countries are trying to improve the conditions of the farming communities, which hopefully will help to prevent fungal infections and other related diseases.

Systemic Mycoses

As the word suggests, "systemic" means the disease can affect the entire system, or body, of the victim. These fungal infections are extremely serious and require immediate medical attention.

Histoplasmosis, for example, is a systemic mycosis that can occur after a person has inhaled a fungus that grows as a mold in contaminated soil or in the droppings of certain birds and bats. The fungus can be breathed into human lungs, where it may grow and multiply. Eventually, the fungus can move to other parts of the body. The person might suffer for many years with the disease, developing a cough and shortness of breath. Ninety percent of all patients with advanced stages of the disease do not survive.

Overall health standards continue to improve around the world. We understand more about germs such as fungi and the diseases they cause. Still, problems such as climate change, disrupted ecosystems, and overcrowding seem to be helping fungi thrive and making it harder to avoid diseases caused by them.

AVOIDING FUNGI IN LOCKER ROOMS

Certain fungi are infamous for living in locker rooms. These damp, sweaty places are a breeding ground for bacteria and fungi, including fungal dermatophytes. Dermatophytes cause the infection commonly known as athlete's foot. When you have athlete's foot, it causes an itchy rash on the feet. It can be scaly and can burn; the skin between the toes may peel. While not life-threatening, it is not a pleasant condition.

The most common way to become infected by dermatophytes is to come into direct contact with a surface contaminated by them. Keeping your feet clean and dry after exercising is a strong defense, but keeping them off locker room floors is also important. Wearing shoes such as flip-flops in public locker rooms and showers protects your feet from coming into contact with the fungus.

If you take steps to prevent it and still come in contact with the fungus, it's not the end of the world! An antifungal cream can treat the symptoms and kill the fungus.

WHAT DO YOU THINK?

1. How can fungi killing plant crops affect human health?
2. How could you identify whether spores or pollen exacerbate your allergies?
3. Why would overcrowding help fungi thrive?

A squeegee tool is great for keeping the shower walls dry. By pushing water droplets down to the drain, you can make it more difficult for fungi to thrive.

CHAPTER 4

HOW CAN WE PREVENT ILLNESS?

While it's established that fungi are essential to our everyday lives, we still want to keep them under control. Controlling their spread, especially in places we inhabit, is the best way to prevent illness caused by them. Fungi feed and thrive in so many habitats that they can be quite difficult to get rid of if we let them spread.

Besides being a health concern, fungi can cause serious damage if allowed to spread indoors. While it's hard to avoid them completely, taking measures to reduce their likelihood and remove them when they begin to grow can keep fungi in check.

Fungi in Homes

Houses and apartments are cozy places for humans and pets, but they also can provide a nice haven for molds and fungi if conditions are right. Any place that is damp tends to attract fungi. Obvious places include the shower, sink, and toilet in the bathroom, as well as the kitchen sink. Because of its size and daily use, the shower tends to be the biggest mold target.

Be sure to clean the shower after you use it. A squeegee works well in removing excess moisture from the shower walls and the shower floor. As often as possible, spray the shower with a disinfectant cleanser and scrub it down to remove dirt and oil.

Although not visible, our bodies release oil and dead skin each time we bathe. The oil can collect in the shower and attract fungi, so it is important to clean the shower just as you clean yourself.

Carpeting is one of the worst culprits for attracting molds and mildew. If the subflooring is not sealed properly, moisture can be absorbed into carpets and rugs resting on top of it. Run your hand over the carpeting to check for dampness. However, even if the carpet is not damp, it can hold spores and support mold growth.

While you may think you're just removing dirt and crumbs when you vacuum, you're also preventing the spread of mold spores.

Be sure to vacuum the carpet regularly to prevent mold growth and the spread of spores. There are many different vacuums on the market. Some work better than others. If you have a lot of carpeting, it is worth investing in a vacuum cleaner with good suction and an airtight filtering system.

Fungi in Vehicles

Cars make attractive homes for fungi. Be sure to vacuum and clean any carpeting, fabric, and other places in the car that might attract molds. It is important to quickly clean up spills in the car before they

While most people clean their houses regularly, sometimes cars get a bit neglected. Still, keeping your car clean is key to mold prevention, especially if you eat or drink in the vehicle.

develop into bigger problems. For example, someone might spill milk in the car over the weekend. If the spill is not properly cleaned up, by the end of the following week that area could produce a rotten smell that could spread to other areas of the car. That is because the milk spoiled and then attracted bacteria and molds to the area.

Fungi in the Air

Any kind of forced air system, such as central heating or air conditioning, can spread mold spores and fungi if not maintained properly. Do what you can to keep vents clean and dry. It then might be necessary to hire professionals to come in and take apart the unit for a more thorough cleaning. At some point, you might have seen a heating company do this. They bring in giant vacuum cleaners that are not practical for homeowners to own. Hoses attached to the vacuums can go deep into areas that you could not reach with a handheld vacuum.

Fungi in Food

Mushrooms are a rare example of a fungus that is good to eat. Mushrooms have a lot of vitamins and minerals that are both healthy and tasty to consume. For other foods, try to prevent mold and fungus by storing them properly. Remember that molds and fungi prefer warm places. Refrigerators and freezers help to prevent fungi from growing because they keep food very cold. For foods that cannot be refrigerated, store them in airtight containers. This too will help to prevent fungi because mold spores in the air will not be able to land on the food.

Refrigerators keep our food fresh and cold, which is key to preventing the growth of fungi.

FUNGI ON YOU

At some point, all of us suffer from cuts, scrapes, bruises, and perhaps even more serious injuries. If possible, try to clean the wounded area right away. Dirt can hold molds and bacteria that could infect the cut. When possible, use an antifungal spray or cream to kill any germs or mold spores that might be present. Finally, cover the wound with a bandage. The bandage will promote healing and protect the injured area.

Basic **hygiene** is always our first line of defense against germs, including fungi, but it is easy for fungi to get the upper hand because they grow and spread quickly. Keeping areas clean and dry that are potential breeding grounds for harmful fungi can make it **inhospitable** for the them.

If a leak or spill occurs, make sure the area is thoroughly dried, and watch for any fungal activity so you can stop it before it takes over. If we take these steps to protect ourselves, fungi remain a minor threat to our health.

Our skin is our immune system's first line of defense. Basic wound care is important in keeping germs, including fungi, out.

FUNGI PROTECTING US

While protecting ourselves against harmful fungi is important, did you know fungi have been key to fighting disease for ages? In fact, even in ancient Egypt, before there was a true understanding of germs, there was a practice of applying moldy breads to wounds. It wasn't until Alexander Fleming's lucky mistake that we started truly using fungi to treat illness, though.

In 1928, Alexander Fleming, a professor of bacteriology, left cultures of staphylococcus, a kind of bacteria, in **petri dishes** when he went on vacation. Upon returning, he noticed that mold had begun to grow on one of the dishes. While this wouldn't be a scientific discovery on its own, he also noticed that the bacteria were not growing around where the mold was.

This "mold juice," later identified as *Penicillium notatum*, was able to kill many different harmful bacteria. Fleming published his findings, which were eventually used about a decade later to create the antibiotic penicillin. This drug, created with a fungus, is still essential today when treating many bacterial infections.

WHAT DO YOU THINK?

1. Besides the bathroom, where else in our homes might fungi thrive?
2. Why are cars such hospitable homes for fungi?
3. How does getting rid of spores prevent the spread of fungi?

GLOSSARY

aflatoxin: A common form of mycotoxin that can cause health problems in animals and humans.

AIDS: Aquired immunodeficiency syndrome, a disease caused by the human immunodeficiency virus (HIV) that interferes with the body's ability to fight infections.

cultivate: To raise or grow, especially on a large scale.

enzyme: A type of protein that creates chemical reactions in the body.

hay fever: A name for seasonal allergic reactions that include symptoms such as sneezing, runny nose, and red, watery, itchy eyes.

hygiene: Practices that help maintain health and prevent disease, especially through cleanliness.

immune system: The body's system of defense against infections.

immunocompromised: Having an impaired immune system.

inhospitable: Describing an environment that is harsh and difficult to live in.

ingest: To consume or take in.

mycotoxin: A potentially dangerous chemical produced by some fungi.

pathogen: A microorganism that can cause disease.

petri dish: A shallow, circular, transparent dish with a flat lid, used in a laboratory to grow microorganisms.

residue: A small amount of something that remains after the main part has gone.

systemic disease: A disease that affects the entire body rather than a single organ or body part.

FIND OUT MORE

BOOKS

Anthony, William. *Foul Fungi*. New York, NY: Enslow Publishing, 2021.

Hawksworth, D. L., and Ester Gaya. *Fungarium*. Somerville, MA: Big Picture Press, 2021.

Owen, Ruth. *Fungi: Colorful Clean-Up Crews*. Kent, UK: Ruby Tuesday Books, 2021.

WEBSITES

Britannica Kids
kids.britannica.com
On the website's home page, there is a search feature that allows you to access articles, images, and videos at different grade levels, including articles on fungus, mold, and yeast, among countless others.

"The Fungus Among Us"
calscomm.cals.cornell.edu/naturalist/Naturalist-Outreach-Fungus.pdf
Created by Cornell University's Naturalist Outreach, this document is a teacher resource guide with a great overview of fungi and additional resources.

Microbiology Society
microbiologysociety.org/why-microbiology-matters/what-is-microbiology/fungi.html
This page includes an overview of fungi along with links to related videos, articles, and podcasts.

ORGANIZATIONS

Association of Medical Microbiology and Infectious Disease Canada (AMMI)
192 Bank Street
Ottawa, ON K2P 1W8
Canada
Email: info@ammi.ca
www.ammi.ca
The AMMI promotes the prevention, diagnosis, and treatment of human infectious diseases through its involvement in education, research, clinical practice, and patient advocacy.

Centers for Disease Control and Prevention (CDC)
1600 Clifton Road
Atlanta, GA 30329
www.cdc.gov
The CDC is the primary U.S. government agency in charge of tracking, treating, and preventing public health threats, including infectious diseases.

National Institutes of Health (NIH)
9000 Rockville Pike
Bethesda, MD 20892
Email: NIHinfo@od.nih.gov
www.nih.gov
The NIH is a research facility in the Washington, D.C., area that is the primary agency of the U.S. government in charge of biomedical and health-related research.

Publisher's note to educators and parents: Our editors have carefully reviewed these websites to ensure that they are suitable for students. Many websites change frequently, however, and we cannot guarantee that a site's future contents will continue to meet our high standards of quality and educational value. Be advised that students should be closely supervised whenever they access the Internet.

INDEX

About the Author

Kathryn Kroe has a master's degree in public health from the University at Buffalo and a bachelor's degree in education from SUNY Buffalo State College. She is passionate about spreading accurate information and helping young people understand issues pertaining to their health and the health of those around them. She currently lives in Lockport, New York, with her husband and two toddlers. She enjoys walking with her family, playing with her Australian shepherd, Zoey, and traveling to see new places.